Engineering Bulletin No 1: Boiler and Furnace Testing

By

Rufus T. Strohm

January 1877

New York

NECESSITY FOR TESTING BOILERS

A boiler test is necessary in order to determine how well the boiler is doing the work expected of it; that is to say, we must find out whether we are wasting coal in making steam and how much this waste may be. Such a test may be made to discover the efficiency of the boiler, or the quantity of water it is evaporating, or the cost of evaporating 1,000 pounds of water.

The United States Fuel Administration recommends that every boiler plant have some means of daily checking the efficiency of the boiler and furnace. The simplest and best way of finding out how efficiently the boiler is working is to make an evaporation test, as described in this bulletin. All the necessary records can be made automatically with suitable instruments, although in many small plants the coal must be weighed on ordinary scales. The efficiency of the furnace can be found by making analyses of the flue gases. (See Bulletin No. 2 of the United States Fuel Administration.)

Too many engineers and firemen have the idea that they are not fitted to make boiler tests. This is altogether wrong. Any man who can weigh water and coal and read steam gages and thermometers is able to do the work required in making a boiler test for evaporation or efficiency. Such a test requires a knowledge of the following:

1. The total weight of coal used.

2. The total weight of water fed to and evaporated by the boiler.

3. The average temperature of the feed water.

4. The average steam pressure in the boiler.

If these four items are known, a series of simple calculations will show how much water is being evaporated per pound of coal, and the efficiency of the boiler and furnace.

To make a test, the following apparatus and instruments are necessary:

1. Scales to weigh the coal.

2. Apparatus to weigh or measure the feed water.

3. Thermometers to take feed-water temperature.

4. Gages to indicate steam pressure.

A boiler test to be of value should extend over a period of at least eight hours. The longer the test the more accurate the results.

For the sake of simplicity, only the essential elements of boiler and furnace testing are treated in this bulletin. For rules covering the refinements for an exhaustive test, the reader is referred to the boiler test code of the American Society of Mechanical Engineers. Copies of this code can be obtained from the secretary, 29 West Thirty-ninth Street, New York City.

WEIGHING THE COAL

The weight of coal used during a test may easily be found by using an ordinary wheelbarrow and a platform scales, arranged as in figure 1. At each side of the scales build an incline with its top level with the top of the platform, but take care not to have either one touch the platform. Set the empty wheelbarrow on the scales, run the movable weight or poise out until it exactly balances the weight of the barrow and lock it in position with the thumbscrew.

Next, put weights on the scale pan *A* to correspond to a net weight of 250 or 300 pounds of coal. Fill the barrow with coal, run it on the scales, and add coal or take off coal until the scales balance. This is easily done by having a small pile of coal *B* beside the scales. If the weights on the scale pan represent, say, 300 pounds, the net weight of coal in the barrow is exactly 300 pounds. This coal is wheeled in front of the boiler and dumped on the clean floor, and the barrow is returned for another load.

Each time the barrow of coal is weighed on the scales and taken to the boiler being tested, a tally mark should be made on a board nailed to the wall beside the scales. Each tally mark represents 300 pounds of coal, since the amount of coal in the barrow is adjusted at each weighing, so that the scales just balance. At the end of the test, therefore, the number of tally marks is multiplied by 300, and the product is the weight of coal used, provided it has all been fired; but if any coal remains in front of the boiler at the close of the test, it must be gathered up and weighed, and its weight must be subtracted from the total weight indicated by the tally marks to get the number of pounds of coal actually fired. You should, of course, start the test with no coal in front of the boiler.

Care must be taken not to forget to make a tally mark each time a barrow of coal is run off the scales. By setting the scales so as to show any net weight, such as 250 or 300 pounds, and making each barrow load exactly this weight, much time is saved, as it is unnecessary to change any of the weights or the position of the rider on the scale beam.

If the coal used in the test is to be analyzed, take a sample of from 4 to 6 pounds from each barrow and throw it into a box near the scales. Do this *before* the coal is weighed. These small amounts from the various barrow loads will then give a fair average sample of the coal used during the test.

The condition of the furnace should be the same at the end of the test period as at the start. Therefore, at the moment the test is begun, observe the thickness of the fuel bed and the condition of the fire. If the fire was cleaned, say, an hour before the test began, see that it is cleaned an hour before the time when the test is scheduled to end. If the coal was fired, say, eight minutes before the test started, the last coal used during the test should be fired eight minutes before the end of the test. The object of these precautions is to insure the same conditions at start and finish, as nearly as possible; otherwise, the coal weighed will not be the same as the coal consumed.

MEASURING THE FEED WATER

The quantity of water fed to the boiler during the test may be found by metering or by weighing. A reliable water meter is recommended for this work. There are a number of good makes, of different types, such as:

1. Venturi meter.

2. Weir or V-notch meters.

3. Diaphragm meters.

4. Displacement meters.

5. Water weighers.

The best form of meter to use in any particular case depends on the local conditions in the plant; but *every plant should be provided with a permanently installed meter of some type*. The displacement form of meter should be used only with cold water, however.

If there is no meter or water weigher in the plant, the feed water used during the test can be measured by the three-barrel arrangement illustrated in figure 2.

Obtain three water-tight barrels, and set two of them close together on a platform directly over the third, leaving about 12 inches above barrel 3 in which to fit the valves *V* and the nipples in the bottoms of barrels 1 and 2. Near the top of each of the barrels 1 and 2 screw a 1-inch overflow pipe *O*.

Run a pipe *P* from the city main or other source of supply above barrels 1 and 2, and put a valve *A* on the pipe leading to each barrel. From barrel 3 run a suction pipe to the feed pump that is to pump water to the boiler to be tested. It is best to have a by-pass from the usual water supply direct to the feed pump, or to another pump connected to the boiler, so that in case of any trouble with the testing barrels, the regular operation of the boiler may be resumed without shutting down.

The next step is to fill barrels 1 and 2 with water until they overflow at O. This water should be of practically the same average temperature as that which is to be used during the test. Barrel 3 should be high enough above the feed pump so that the pump will handle hot water. Put barrel 3 on a scales, before connecting it to the feed pump, and weigh it. Then let the water from barrel 1 run into barrel 3, and weigh again. The second weight minus the first weight is the net weight of water run in from barrel 1 and is the weight of water contained in barrel 1 when filled to the overflow. The weight of water in barrel 2 when it is filled to the overflow can be found in like manner. Mark these weights down.

When the net weights are found and barrel 3 is removed from the scales and connected to the feed pump, the apparatus is ready to begin the test. Start with the level of the water about 1 foot below the top of the barrel 3, and drive a nail into the barrel to mark this level. When the test is finished, the level should be brought to the same point, so that the water that has passed through barrels 1 and 2 will accurately represent the weight of water fed to the boiler during the test.

When the test is to begin, stop the feed pump and tie a string around the gage glass on the boiler to mark the height of the water level in the boiler. Then start the pump connected to barrel 3. Fill barrels 1 and 2 up to the overflow before the test is started. Then open the valve V on barrel 1 and let the water run into barrel 3 as fast as the feed pump draws water from barrel 3. When barrel 1 is emptied close its valve V and open its valve A so as to refill it.

While barrel 1 is filling empty barrel 2 into barrel 3 in the same way, and continue to fill and empty barrels 1 and 2 alternately. In this way barrel 3 will be kept supplied with water that has been measured in barrels 1 and 2, the net weights of which were found before the test began. Keep a separate tally of the number of times each of the barrels 1 and 2 is emptied into barrel 3. At the end of the test the number of tallies for each barrel multiplied by the weight of the water that barrel will hold will be the weight of water measured in that barrel. The sum of these weights for barrels 1 and 2 will be the weight of water used in the test.

With a three-barrel arrangement like this, water can be weighed rapidly enough to supply 300 boiler horsepower.

Before starting a test make sure that there is no chance for water to leak into or out of the boiler. See that the blow-off is tight, that there is no drip from gage cocks, and that the feed-line connections are tight, so that all the water fed to the boiler will represent accurately the amount evaporated during the test.

If a meter is used instead of the three-barrel method, make absolutely sure that the meter is correct, as the accuracy of the test depends on the accuracy with which the water measurements are made. *After a meter is installed, test it to see that it operates correctly under the plant conditions.*

The water level in the boiler should be the same at the end of the test as at the beginning. As the time for stopping the test draws near, therefore, try to bring the conditions the same as at the start. Do not, however, run the feed pump rapidly in the last few minutes for the test in order to obtain the same water level. If there is a slight difference in level, calculate the weight of water it represents and make the necessary correction to the total weight of water fed.

TEMPERATURE OF FEED WATER

Every plant should have a thermometer on the feed line, so as to find the temperature of the feed water. Preferably, this thermometer should be of the recording type. If such a form of thermometer is used during the test, it is unnecessary to take the feed temperature at stated intervals, as the record will show the varying temperatures, and so the average feed temperature during the test can easily be found.

If there is no thermometer in the feed line, take the feed-water temperature by means of a thermometer hung in barrel 3 (figure 2) by a hook over the edge of the barrel. Read this thermometer every half hour during the test if the feed-water temperature is fairly uniform; but if it varies considerably, read the thermometer every 15 minutes. The object is to obtain the average feed-water temperature during the test period. Therefore, mark down the tempera tures as read at the stated intervals. At the close of the test add the readings and divide their sum by the number of readings and you will have the average temperature of the feed water.

STEAM PRESSURE

Every boiler is fitted with a steam gage by which the pressure is indicated. It is important that the pressure gage be accurate. What is wanted in a test is the average pressure of the steam in the boiler, therefore, observe the pressure at regular intervals, just as with the feed-water temperature, and mark down these gage readings. The sum of the readings divided by the number of readings taken will be the average steam pressure during the test.

A recording steam gage is best and makes its own readings.

WORKING UP THE TEST

After the boiler test has been made, so as to find the weight of coal burned, weight of feed water used, feed-water temperature and steam pressure, the efficiency, the horsepower, and the economy must be obtained by calculation from the test results. The process of figuring the desired results from the test data is called "working up the test."

To illustrate the method used in finding the efficiency, etc., suppose that the data obtained from the test are as follows:

Length of test.....................................	hours	10
Total weight of coal fired....................	pounds	5,000
Total weight of water evaporated.......	do.	35,000
Average temperature of feed water....	°F	180
Average steam pressure, gage............ pounds per square inch		100

The efficiency of any process is always a comparison, or ratio, of the output to the input. In the case of a steam boiler the efficiency is the percentage of the heat supplied in the coal that is usefully employed in making steam. The output of the steam boiler is the heat represented by the quantity of water evaporated by a pound of coal, taking into account the feed temperature and the steam pressure, and input is the amount of heat contained in a pound of the coal used. The efficiency of the boiler is the output divided by the input.

The heat contained in a pound of coal is called the "calorific value" or "heating value" of the coal. It can be found by taking a fair average sample of the coal used during the test, as explained in connection with weighing the coal, and sending the sample to a chemist, who will make a calorimeter test to determine its heating value.

At the end of the test the sample fuel should be spread out on a clean floor and all lumps broken up, so that no pieces are larger than 2 inches maximum diameter. Then the gross sample should be very thoroughly mixed by shoveling, after which it should be spread out in the form of a square of uniform depth and quartered down until[9] a final average sample is obtained for shipment to a competent chemist, experienced in fuel analysis. (See Bureau of Mines Technical Paper No. 133.)

About 2 quarts of the chemist's sample should be put in air-tight tins or jars for the determination of moisture; the balance of the sample (the total weight of which should be from 10 to 50 pounds, depending on the total weight of coal used in the test) may be packed in a wooden box lined with paper to prevent splinters from mingling with the sample. A duplicate coal sample should be kept at the plant to be used in case of loss of the sample sent to the chemist.

The Bureau of Mines has published a bulletin or pamphlet giving the analyses and heating values of the various kinds and grades of coal from all parts of the United States. (Bureau of Mines Bulletin No. 22.) This bulletin can be used to learn the approximate heating value of the coal. Simply find out what district the coal used in the test came from, and its grade, and then refer to the bulletin to obtain the heating value of the coal. If a chemist can be obtained to make a heat test, however, it is better to use the heating value he determines.

Suppose that during the test the coal used was run-of-mine bituminous having a heating value of 13,500 B. t. u. Every pound of coal fired, then, carried into the furnace 13,500 heat units, and this value therefore is the *input* to be used in calculating the boiler efficiency.

During the test 5,000 pounds of coal was fired and 35,000 pounds of water was fed and evaporated. This means that $35,000 \div 5,000 = 7$ pounds of water was evaporated per pound of coal burned. This is the "actual evaporation," and the heat required to evaporate this 7 pounds of water is the output to be used in calculating the efficiency.

Every fireman knows that it takes more coal, and therefore more heat, to make steam with cold feed water than with hot feed water; also, that it is somewhat easier to make steam at a low pressure than at a high pressure.

So it is plain that the heat required to evaporate 7 pounds of water into steam depends on two things, namely, (1) the temperature of the feed water and (2) the pressure of the steam in the boiler. From the data of the test, both the average feed-water temperature and the average steam pressure are known, and so it is a simple matter to find out the amount of heat needed to evaporate 7 pounds of water from the average temperature to steam at the average pressure.

A pound of water at 212° F. must have 970.4 B. t. u. added to it to become a pound of steam at 212° F., or zero gage pressure. This value, 970.4 B. t. u., is called the latent heat of steam at atmospheric pressure, or the heat "from and at 212° F." It is the heat required to change a pound of water *from* 212° F. to steam *at* 212° F.,[10] and is used by engineers as a standard by which to compare the evaporation of different boilers.

In a boiler test the temperature of the feed water is usually something less than 212° F., and the steam pressure is commonly higher than zero, gage. In the test outlined previously, the feed-water temperature was 180° F. and the pressure was 100 pounds per square inch, gage. It must be clear, then, that the amount of heat required to change a pound of water at 180° to steam at 100 pounds gage pressure is not the same as to make a pound of steam from and at 212° F.

To make allowance for the differences in temperature and pressure, the actual evaporation must be multiplied by a number called the "factor of evaporation." The factor of evaporation has a certain value corresponding to every feed-water temperature and boiler pressure, and the values of this factor are given in the accompanying table. Along the top of the table are given the gage pressures of the steam. In the columns at the sides of the table are given the feed-water temperatures. To find the factor of evaporation for a given set of conditions, locate the gage pressure at the top of the table and follow down that column to the horizontal line on which the feed-water temperature is located. The value in this column and on the horizontal line thus found is the factor of evaporation required. If the feed water has a temperature greater than 212° F., obtain the proper factor of evaporation from the Marks and Davis steam tables.

Take the data of the test, for example. The average steam pressure is 100 pounds, gage. The average feed-water temperature is 180° F. So, in the table locate the column headed 100 and follow down this column to the line having 180 at the ends, and the value where the column and the line cross is 1.0727, which is the factor of evaporation for a feed-water temperature of 180° F. and a steam pressure of 100 pounds, gage.

This factor, 1.0727, indicates that to change a pound of water at 180° F. to steam at 100 pounds requires 1.0727 times as much heat as to change a pound of water at 212° F. to steam at atmospheric pressure. In other words, the heat used in producing an actual evaporation of 7 pounds under the test conditions would have evaporated 7 × 1.0727 = 7.5 pounds from and at 212° F. Hence, 7.5 pounds is called the "equivalent evaporation from and at 212° F." per pound of coal used.

As already stated, it takes 970.4 B. t. u. to make a pound of steam from and at 212° F. Then to make 7.5 pounds there would be required 7.5 × 970.4 = 7,278 B. t. u. This is the amount of heat required to change 7.5 pounds of water at 212° F. to steam at zero gage pressure, but it is also the heat required to change 7 pounds[11] of water at 180° F. to steam at 100 pounds gage pressure, because 7.5 pounds from and at 212° F. is equivalent to 7 pounds from 180° F. to steam at 100 pounds. Therefore, the 7,278 B. t. u. is the amount of heat usefully employed in making steam per pound of coal fired, and so it is the *output*. Accordingly, the efficiency of the boiler is—

$$\sim \text{Efficiency} = \frac{\text{Output}}{\text{Input}} = \frac{7{,}278}{13{,}500} = 0.54, \text{ nearly.}$$

In other words, the efficiency of the boiler is 0.54, or 54 per cent, which means that only a little more than half of the heat in the coal is usefully employed in making steam.

The chart shown in figure 3 is given to save the work of figuring the efficiency. If the equivalent evaporation per pound of coal is calculated and the heating value of the coal is known, the boiler efficiency may be found directly from the chart. At the left-hand side locate the point corresponding to the equivalent evaporation and at the bottom locate the point corresponding to the heating value of the coal. Follow the horizontal and vertical lines from these two points until they cross, and note the diagonal line that is nearest to the crossing point. The figures marked on the diagonal line indicate the boiler efficiency.

Take the case just worked out, for example. The equivalent evaporation is 7.5 pounds and the heating value of the fuel is 13,500 B. t. u. At the left of the chart locate the point 7.5 midway between 7 and 8 and at the bottom locate the point 13,500 midway between 13,000 and 14,000. Then follow the horizontal and vertical lines from these two points until they cross, as indicated by the dotted lines. The crossing point lies on the diagonal corresponding to 54, and so the efficiency is 54 per cent.

BOILER HORSEPOWER OR CAPACITY.

The capacity of a boiler is usually stated in boiler horsepower. A boiler horsepower means the evaporation of 34.5 pounds of water per hour from and at 212° F. Therefore, to find the boiler horsepower developed during a test, calculate the evaporation from and at 212° F. per hour and divide it by 34.5.

Take the test previously mentioned, for example. The evaporation from and at 212° F. or the equivalent evaporation, was 7.5 pounds of water per pound of coal. The weight of coal burned per hour was 5,000 ÷ 10 = 500 pounds. Then the equivalent evaporation was 7.5 × 500 = 3,750 pounds per hour. According to the foregoing definition of a boiler horsepower, then—

$$\text{Boiler horsepower} = \frac{3{,}750}{34.5} = 109.$$

[12]

The "rated horsepower" of a boiler, or the "builders' rating," is the number of square feet of heating surface in the boiler divided by a number. In the case of stationary boilers this number is 10 or 12, but 10 is very commonly taken as the amount of heating surface per horsepower. Assuming this value and assuming further that the boiler tested had 1,500 square feet of heating surface, its rated horsepower would be 1,500 ÷ 10 = 150 boiler horsepower.

It is often desirable to know what per cent of the rated capacity is developed in a test. This is found by dividing the horsepower developed during the test by the builders' rating. In the case of the boiler tested, 109 horsepower was developed. The percentage of rated capacity developed, therefore, was 109 ÷ 150 = 0.73, or 73 per cent.

HEATING SURFACE.

The heating surface of a boiler is the surface of metal exposed to the fire or hot gases on one side and to water on the other side. Thus, the internal surface of the tubes of a fire-tube boiler is the heating surface of the tubes, but the outside surface of the tubes of a water-tube boiler is the heating surface of those tubes. In addition to the tubes, all other surfaces which have hot gases on one side and water on the other must be taken into account. For instance, in a fire-tube boiler from one-half to two-thirds of the shell (depending on how the boiler is set) acts as heating surface. In addition to this, the surface presented by both heads, below the water level, has to be computed. The heating surface of each head is equal to two-thirds its area minus the total area of the holes cut away to receive the tubes.

COST OF EVAPORATION.

The cost of evaporation is usually stated as the cost of fuel required to evaporate 1,000 pounds of water from and at 212° F. To find it, multiply the price of coal per ton by 1,000 and divide the result by the product of the equivalent evaporation per pound of coal and the number of pounds in a ton.

Suppose that the cost of the coal used in the foregoing test was $3.60 per ton of 2,000 pounds. The equivalent evaporation per pound of coal was 7.5 pounds. Therefore the cost of evaporating 1,000 pounds of water from 180° F. to steam at 100-pound gage, is—

$$\frac{\$3.60 \times 1,000}{7.5 \times 2,000} = \$0.24, \text{ or 24 cents.}$$

[13]

TABLE OF TEST RESULTS.

After the test has been made and properly worked up, as heretofore described, collect all the results of the test on one sheet, so that they can be kept in convenient form for reference and for comparison with later tests. A brief form of arranging the results is as follows:

		May 20, 1918
1. Date of test..		
2. Duration of test...	hours	10
3. Weight of coal used...	pounds	5,000
4. Weight of water fed and evaporated...	do.	35,000
5. Average steam pressure, gauge....................................	do.	100

6. Average feed-water temperature...	°F.	180
7. Factor of evaporation...		1.0727
8. Equivalent evaporation from and at 212° F.................................	pounds	37,545

EFFICIENCY.

9. Efficiency of boiler and furnace...	per cent	54

CAPACITY.

10. Boiler horsepower developed..		109
11. Builders' rated horsepower..		150
12. Percentage of rated horsepower developed...............................	per cent	73

ECONOMIC RESULTS.

13. Actual evaporation per pound of coal...	pounds	7
14. Equivalent evaporation from and at 212° F.................................	pounds	7.5
per pound of coal as fired,		
15. Cost of coal per ton (2,000 pounds)..		$3.60
16. Cost of coal to evaporate 1,000 pounds from and at 212° F.....		$0.24

HOW TO USE THE TEST RESULTS.

The object of working up a test is to obtain a clear idea as to the efficiency of operation of the boiler or its operating cost. Consequently, after the calculations have been made, they should be used as a basis for study with the idea of improving the boiler performance.

Take the matter of boiler efficiency, for example, as found from the test mentioned. Its value was 54 per cent. This is altogether too low and indicates wasteful operation. The efficiency of a hand-fired boiler ought not to be less than 65 per cent, and it can be increased to 70 per cent by careful management under good conditions.

The chart in figure 3 can be used to indicate the evaporation that should be obtained in order to reach a desired efficiency. Suppose, for example, that it is desired to know how much water per pound of coal must be evaporated to produce a boiler efficiency of 65 per cent with coal having a heating value of 13,500 B. t. u. per pound.

Locate 13,500 at the bottom of the chart, follow the vertical line until it meets the diagonal marked 65 per cent, and then from this point follow the horizontal line to the left-hand edge, where the[14] figure 9 is found. This means that the equivalent evaporation from and by 212° F. per pound of coal must be 9 pounds of water. If the steam pressure is 100 pounds gauge, and the feed-water temperature is 180° F. the factor of evaporation is 1.0727, then the actual evaporation must be 9 ÷ 1.0727 = 8.36 pounds per pound of coal. In other words, to increase the efficiency from 54 per cent to 65 per cent under the same conditions of pressure and feed-water temperature, it would be necessary to increase the actual evaporation from 7 pounds to 8.36 pounds. This would mean practically 20 per cent more steam from the same weight of coal used.

[15]

How to do this will require some study and experimenting on the part of the fireman or engineer. The three most common reasons for low-boiler efficiency are (1) excess air, (2) dirty heating surfaces, and (3) loss of coal through the grates. *The first of these items is the most important of the three.* In most cases the greatest preventable waste of coal in a boiler plant is directly due to excess air. Excess air simply means the amount of air which gets into the furnace and boiler which is not needed for completing the combustion of the coal. Very often twice as much air is admitted to the boiler setting as is required. This extra or excess air is heated and carries heat out through the chimney instead of heating the water in the boiler to make steam. There are two ways in which this excess air gets into the furnace and boiler setting. First, by a combination of bad regulation of drafts and firing. The chances are your uptake damper is too wide open. Try closing it a little. Then, there may be holes in the fire. Keep these covered. The second way excess air occurs is by leakage through the boiler setting, through cracks in the brickwork, leaks around the frames and edges of cleaning doors, and holes around the blow-off pipes. There are also other places where such air can leak in.

Take a torch or candle and go over the entire surface of your boiler setting—front, back, sides, and top. Where the flame of the torch is drawn inward there is an air leak. Plaster up all air leaks and repair the brickwork around door frames where necessary. You should go over your boiler for air leaks once a month.

In regard to best methods of firing soft coal, see Technical Paper No. 80 of the Bureau of Mines, which may be obtained from your State Fuel Administrator.

Dirty heating surfaces cause low efficiency because they prevent the heat in the hot gases from getting through into the water. Therefore, keep the shell and tubes free of soot on one side and scale on the other. Soot may be removed by the daily use of blowers, scrapers, and cleaners. The problem of scale and pure feed water is a big one and should be taken up with proper authorities on the subject.

There are many things that may be done to increase the efficiency of the boiler and to save coal. For convenience a number of these points are grouped in the following list:

[16]

WHAT TO DO.	WHY.
1. Close up all leaks in the boiler setting.	To prevent waste of heat due to excess air admitted.
2. Keep shell and tubes free from soot and scale.	To allow the heat to pass easily into the water.
3. Use grates suited to the fuel to be burned.	To prevent loss of unburnt coal through air spaces.
4. Fire often, and little at a time.	To obtain uniform conditions and better combustion.
5. Cover all thin spots and keep fire bed level.	To prevent burning holes in bed and admitting excess air.
6. Do not allow clinkers to form on side or bridge walls.	Because they reduce the effective area of the grate.
7. Keep the ash pit free from ashes and hot clinkers.	To prevent warping and burning out of the grates.
8. Do not stir the fire except when necessary.	Because stirring causes clinker and is likely to waste coal.
9. Use damper and not ash-pit doors to control draft.	Because less excess air is admitted by so doing.
10. See that steam pipes and valves are tight.	Because steam leaks waste heat and therefore coal.
11. Keep blow-off valves tight.	Because leaks of hot water waste coal.
12. Cover steam pipes and the tops of boilers.	To prevent radiation and loss of heat.

Make a boiler test under the conditions of operation as they now exist in your plant. Then make all possible improvements as suggested in this bulletin, make another test afterwards and note the increase in the equivalent evaporation per pound of coal used.

Remember that the *firing line* in the boiler room can be just as patriotic and helpful as the *firing line* at the front.

[17]

Table of factors of evaporation.

Feed temperature , °F.	Steam pressure in pounds per square inch, gauge.							
	30	50	70	80	90	100	110	120
32.......	1.2073	1.2144	1.2195	1.2216	1.2234	1.2251	1.2266	1.2279
35.......	1.2042	1.2113	1.2164	1.2184	1.2203	1.2219	1.2235	1.2248
38.......	1.2011	1.2082	1.2133	1.2153	1.2172	1.2188	1.2204	1.2217
41.......	1.1980	1.2051	1.2102	1.2122	1.2141	1.2157	1.2173	1.2186
44.......	1.1949	1.2020	1.2071	1.2091	1.2110	1.2126	1.2142	1.2155
47.......	1.1918	1.1989	1.2040	1.2060	1.2079	1.2095	1.2111	1.2124
50.......	1.1887	1.1958	1.2009	1.2029	1.2048	1.2064	1.2080	1.2093
53.......	1.1856	1.1927	1.1978	1.1998	1.2017	1.2033	1.2049	1.2062

56.......	1.1825	1.1896	1.1947	1.1967	1.1986	1.2002	1.2018	1.2031
59.......	1.1794	1.1865	1.1916	1.1937	1.1955	1.1972	1.1987	1.2000
62.......	1.1763	1.1835	1.1885	1.1906	1.1924	1.1941	1.1956	1.1970
65.......	1.1733	1.1804	1.1854	1.1875	1.1893	1.1910	1.1925	1.1939
68.......	1.1702	1.1773	1.1823	1.1844	1.1862	1.1879	1.1894	1.1908
71.......	1.1671	1.1742	1.1792	1.1813	1.1832	1.1848	1.1864	1.1877
74.......	1.1640	1.1711	1.1762	1.1782	1.1801	1.1817	1.1833	1.1846
77.......	1.1609	1.1680	1.1731	1.1751	1.1770	1.1786	1.1802	1.1815
80.......	1.1578	1.1650	1.1700	1.1721	1.1739	1.1756	1.1771	1.1785
83.......	1.1548	1.1619	1.1669	1.1690	1.1708	1.1725	1.1740	1.1754
86.......	1.1518	1.1588	1.1638	1.1659	1.1678	1.1694	1.1710	1.1723
89.......	1.1486	1.1557	1.1608	1.1628	1.1647	1.1663	1.1679	1.1692
92.......	1.1455	1.1526	1.1577	1.1597	1.1616	1.1632	1.1648	1.1661
95.......	1.1424	1.1495	1.1546	1.1566	1.1585	1.1602	1.1617	1.1630
98.......	1.1393	1.1465	1.1515	1.1536	1.1554	1.1571	1.1586	1.1600

101.......	1.1363	1.1434	1.1484	1.1505	1.1523	1.1540	1.1555	1.1569
104.......	1.1332	1.1403	1.1453	1.1474	1.1492	1.1509	1.1525	1.1538
107.......	1.1301	1.1372	1.1423	1.1443	1.1462	1.1478	1.1494	1.1507
110.......	1.1270	1.1341	1.1392	1.1412	1.1431	1.1447	1.1463	1.1476
113.......	1.1239	1.1310	1.1360	1.1382	1.1400	1.1417	1.1432	1.1445
116.......	1.1209	1.1280	1.1330	1.1351	1.1369	1.1386	1.1401	1.1415
119.......	1.1178	1.1249	1.1299	1.1320	1.1339	1.1355	1.1370	1.1384
122.......	1.1147	1.1218	1.1269	1.1289	1.1308	1.1324	1.1340	1.1353
125.......	1.1116	1.1187	1.1238	1.1258	1.1277	1.1293	1.1309	1.1322
128.......	1.1085	1.1156	1.1207	1.1227	1.1246	1.1262	1.1278	1.1291
131.......	1.1054	1.1125	1.1176	1.1197	1.1215	1.1232	1.1247	1.1260
134.......	1.1023	1.1095	1.1145	1.1166	1.1184	1.1201	1.1216	1.1230
137.......	1.0993	1.1064	1.1114	1.1135	1.1153	1.1170	1.1185	1.1199
140.......	1.0962	1.1033	1.1083	1.1104	1.1123	1.1139	1.1154	1.1168
143.......	1.0931	1.1002	1.1052	1.1073	1.1092	1.1108	1.1124	1.1137

146.......	1.0900	1.0971	1.1022	1.1042	1.1061	1.1077	1.1093	1.1106
149.......	1.0869	1.0940	1.0991	1.1011	1.1030	1.1046	1.1062	1.1075
152.......	1.0838	1.0909	1.0960	1.0980	1.0999	1.1015	1.1031	1.1044
155.......	1.0807	1.0878	1.0929	1.0950	1.0968	1.0985	1.1000	1.1013
158.......	1.0776	1.0847	1.0898	1.0919	1.0937	1.0954	1.0969	1.0982
161.......	1.0745	1.0817	1.0867	1.0888	1.0906	1.0923	1.0938	1.0952
164.......	1.0715	1.0786	1.0836	1.0857	1.0875	1.0892	1.0907	1.0921
167.......	1.0684	1.0755	1.0805	1.0826	1.0844	1.0861	1.0876	1.0890
170.......	1.0653	1.0724	1.0774	1.0795	1.0813	1.0830	1.0845	1.0859
172.......	1.0632	1.0703	1.0754	1.0774	1.0793	1.0809	1.0825	1.0838
174.......	1.0611	1.0683	1.0733	1.0754	1.0772	1.0789	1.0804	1.0817
176.......	1.0591	1.0662	1.0712	1.0733	1.0752	1.0768	1.0783	1.0797
178.......	1.0570	1.0641	1.0692	1.0712	1.0731	1.0747	1.0763	1.0776
180.......	1.0549	1.0621	1.0671	1.0692	1.0710	1.0727	1.0742	1.0756
182.......	1.0529	1.0600	1.0650	1.0671	1.0690	1.0706	1.0721	1.0735

184.......	1.0508	1.0579	1.0630	1.0650	1.0669	1.0685	1.0701	1.0714
186.......	1.0488	1.0559	1.0609	1.0630	1.0648	1.0665	1.0680	1.0694
188.......	1.0467	1.0538	1.0588	1.0609	1.0628	1.0644	1.0660	1.0673
190.......	1.0446	1.0517	1.0568	1.0588	1.0607	1.0623	1.0639	1.0652
192.......	1.0425	1.0497	1.0547	1.0568	1.0586	1.0603	1.0618	1.0632
194.......	1.0405	1.0476	1.0526	1.0547	1.0566	1.0582	1.0597	1.0611
196.......	1.0384	1.0455	1.0506	1.0526	1.0545	1.0561	1.0577	1.0590
198.......	1.0363	1.0435	1.0485	1.0506	1.0524	1.0541	1.0556	1.0570
200.......	1.0343	1.0414	1.0464	1.0485	1.0504	1.0520	1.0535	1.0549
202.......	1.0322	1.0393	1.0444	1.0464	1.0483	1.0499	1.0515	1.0528
204.......	1.0301	1.0372	1.0423	1.0444	1.0462	1.0479	1.0494	1.0507
206.......	1.0281	1.0352	1.0402	1.0423	1.0441	1.0458	1.0473	1.0487
208.......	1.0260	1.0331	1.0381	1.0402	1.0421	1.0437	1.0453	1.0466
210.......	1.0239	1.0310	1.0361	1.0381	1.0400	1.0416	1.0432	1.0445
212.......	1.0218	1.0290	1.0340	1.0361	1.0379	1.0396	1.0411	1.0425

*Table of factors of evaporation—***Concluded.**

Feed temperature , °F.	Steam pressure in pounds per square inch, gauge.							
	130	140	150	160	170	180	190	200
32.......	1.2292	1.2304	1.2315	1.2324	1.2333	1.2342	1.2351	1.2358
35.......	1.2261	1.2273	1.2283	1.2293	1.2302	1.2311	1.2320	1.2327
38.......	1.2230	1.2242	1.2252	1.2262	1.2271	1.2280	1.2288	1.2296
41.......	1.2199	1.2211	1.2221	1.2231	1.2240	1.2249	1.2257	1.2265
44.......	1.2168	1.2180	1.2190	1.2200	1.2209	1.2218	1.2226	1.2234
47.......	1.2137	1.2149	1.2159	1.2168	1.2178	1.2187	1.2195	1.2202
50.......	1.2106	1.2118	1.2128	1.2137	1.2147	1.2156	1.2164	1.2171
53.......	1.2075	1.2087	1.2097	1.2107	1.2116	1.2125	1.2133	1.2141
56.......	1.2044	1.2056	1.2066	1.2076	1.2085	1.2094	1.2102	1.2110
59.......	1.2013	1.2025	1.2035	1.2045	1.2054	1.2063	1.2072	1.2079
62.......	1.1982	1.1994	1.2005	1.2014	1.2023	1.2032	1.2041	1.2048

65.......	1.1951	1.1963	1.1974	1.1983	1.1992	1.2002	1.2010	1.2017
68.......	1.1920	1.1933	1.1943	1.1952	1.1961	1.1971	1.1979	1.1986
71.......	1.1889	1.1902	1.1912	1.1921	1.1931	1.1940	1.1948	1.1955
74.......	1.1859	1.1871	1.1881	1.1890	1.1900	1.1909	1.1917	1.1924
77.......	1.1828	1.1840	1.1850	1.1860	1.1869	1.1878	1.1886	1.1894
80.......	1.1797	1.1809	1.1820	1.1829	1.1838	1.1847	1.1856	1.1863
83.......	1.1766	1.1778	1.1789	1.1798	1.1807	1.1817	1.1825	1.1832
86.......	1.1735	1.1748	1.1758	1.1767	1.1776	1.1786	1.1794	1.1801
89.......	1.1704	1.1717	1.1727	1.1736	1.1746	1.1755	1.1763	1.1770
92.......	1.1674	1.1686	1.1696	1.1705	1.1715	1.1724	1.1732	1.1739
95.......	1.1643	1.1655	1.1665	1.1675	1.1684	1.1693	1.1701	1.1709
98.......	1.1612	1.1624	1.1635	1.1644	1.1653	1.1662	1.1671	1.1678
101.......	1.1581	1.1593	1.1604	1.1613	1.1622	1.1632	1.1640	1.1647
104.......	1.1550	1.1563	1.1573	1.1582	1.1592	1.1601	1.1609	1.1616
107.......	1.1519	1.1532	1.1542	1.1551	1.1561	1.1570	1.1578	1.1585

110.......	1.1489	1.1501	1.1511	1.1521	1.1530	1.1539	1.1547	1.1555
113.......	1.1458	1.1470	1.1481	1.1490	1.1499	1.1508	1.1515	1.1524
116.......	1.1427	1.1439	1.1450	1.1459	1.1468	1.1478	1.1486	1.1493
119.......	1.1396	1.1409	1.1419	1.1428	1.1437	1.1447	1.1455	1.1462
122.......	1.1365	1.1378	1.1388	1.1397	1.1407	1.1416	1.1424	1.1431
125.......	1.1335	1.1347	1.1357	1.1366	1.1376	1.1385	1.1393	1.1400
128.......	1.1304	1.1316	1.1326	1.1336	1.1345	1.1354	1.1362	1.1370
131.......	1.1273	1.1285	1.1295	1.1305	1.1314	1.1323	1.1332	1.1339
134.......	1.1242	1.1254	1.1265	1.1274	1.1283	1.1292	1.1301	1.1308
137.......	1.1211	1.1224	1.1234	1.1243	1.1252	1.1262	1.1270	1.1277
140.......	1.1180	1.1193	1.1203	1.1212	1.1221	1.1231	1.1239	1.1246
143.......	1.1149	1.1162	1.1172	1.1181	1.1191	1.1200	1.1208	1.1215
146.......	1.1119	1.1131	1.1141	1.1150	1.1160	1.1169	1.1177	1.1184
149.......	1.1088	1.1100	1.1110	1.1120	1.1129	1.1138	1.1146	1.1154
152.......	1.1057	1.1069	1.1079	1.1089	1.1098	1.1107	1.1115	1.1123

155.......	1.1026	1.1038	1.1048	1.1058	1.1067	1.1076	1.1085	1.1092
158.......	1.0995	1.1007	1.1018	1.1027	1.1036	1.1045	1.1054	1.1061
161.......	1.0964	1.0976	1.0987	1.0996	1.1005	1.1014	1.1023	1.1030
164.......	1.0933	1.0945	1.0956	1.0965	1.0974	1.0984	1.0992	1.0999
167.......	1.0902	1.0914	1.0925	1.0934	1.0943	1.0953	1.0961	1.0968
170.......	1.0871	1.0883	1.0894	1.0903	1.0912	1.0922	1.0930	1.0937
172.......	1.0850	1.0863	1.0873	1.0882	1.0892	1.0901	1.0909	1.0916
174.......	1.0830	1.0842	1.0853	1.0862	1.0871	1.0880	1.0889	1.0896
176.......	1.0809	1.0822	1.0832	1.0841	1.0850	1.0860	1.0868	1.0875
178.......	1.0789	1.0801	1.0811	1.0820	1.0830	1.0839	1.0847	1.0854
180.......	1.0768	1.0780	1.0791	1.0800	1.0809	1.0818	1.0827	1.0834
182.......	1.0747	1.0760	1.0770	1.0779	1.0788	1.0798	1.0806	1.0813
184.......	1.0727	1.0739	1.0749	1.0759	1.0768	1.0777	1.0785	1.0793
186.......	1.0706	1.0718	1.0729	1.0738	1.0747	1.0756	1.0765	1.0772
188.......	1.0685	1.0698	1.0708	1.0717	1.0727	1.0736	1.0744	1.0751

190.......	1.066 5	1.067 7	1.068 7	1.069 7	1.070 6	1.071 5	1.072 3	1.073 1
192.......	1.064 4	1.065 6	1.066 7	1.067 6	1.068 5	1.069 4	1.070 3	1.071 0
194.......	1.062 3	1.063 6	1.064 6	1.065 5	1.066 4	1.067 4	1.068 2	1.068 9
196.......	1.060 3	1.061 5	1.062 5	1.063 5	1.064 4	1.065 3	1.066 1	1.066 9
198.......	1.058 2	1.059 4	1.060 5	1.061 4	1.062 3	1.063 2	1.064 1	1.064 8
200.......	1.056 1	1.057 4	1.058 4	1.059 3	1.060 2	1.061 2	1.062 0	1.062 7
202.......	1.054 1	1.055 3	1.056 3	1.057 2	1.058 2	1.059 1	1.059 9	1.060 6
204.......	1.052 0	1.053 2	1.054 2	1.055 2	1.056 1	1.057 0	1.057 9	1.058 6
206.......	1.049 9	1.051 1	1.052 2	1.053 1	1.054 0	1.055 0	1.055 8	1.056 5
208.......	1.047 8	1.049 1	1.050 1	1.051 0	1.052 0	1.052 9	1.053 7	1.054 4
210.......	1.045 8	1.047 0	1.048 0	1.049 0	1.049 9	1.050 8	1.051 6	1.052 4
212.......	1.043 7	1.044 9	1.046 0	1.046 9	1.047 8	1.048 7	1.049 6	1.050 3

PUBLICATIONS ON THE UTILIZATION OF COAL AND LIGNITE.

A limited supply of the following publications of the Bureau of Mines has been printed and is available for free distribution until the edition is exhausted. Requests for all publications can not be granted, and to insure equitable distribution applicants are requested to limit their selection to publications that may be of especial interest to them. Requests for publications should be addressed to the Director, Bureau of Mines.

The Bureau of Mines issues a list showing all its publications available for free distribution, as well as those obtainable only from the Superintendent of Documents, Government Printing Office, on payment of the price of printing. Interested persons should apply to the Director, Bureau of Mines, for a copy of the latest list.

PUBLICATIONS AVAILABLE FOR FREE DISTRIBUTION.

Bulletin 58. Fuel briquetting investigations, July, 1904, to July, 1912, by C. A. Wright. 1913. 277 pp., 21 pls., 3 figs.

Bulletin 76. United States coals available for export trade, by Van. H. Manning. 1914. 15 pp., 1 pl.

Bulletin 85. Analyses of mine and car samples of coal collected in the fiscal years 1911 to 1913, by A. C. Fieldner, H. I. Smith, A. H. Fay, and Samuel Sanford. 1914. 444 pp., 2 figs.

Bulletin 89. Economic methods of utilizing western lignites, by E. J. Babcock. 1915. 74 pp., 5 pls., 5 figs.

Bulletin 119. Analyses of coals purchased by the Government during the fiscal years 1908-1915, by G. S. Pope. 1916. 118 pp.

Bulletin 135. Combustion of coal and design of furnaces, by Henry Kreisinger, C. E. Augustine, and F. K. Ovitz. 1917. 144 pp., 1 pl., 45 figs.

Bulletin 136. Deterioration in the heating value of coal during storage, by H. C. Porter and F. K. Ovitz. 1917. 38 pp., 7 pls.

Bulletin 138. Coking of Illinois coals, by F. K. Ovitz. 1917. 71 pp., 11 pls. 1 fig.

Technical Paper 34. Experiments with furnaces for a hand-fired return tubular boiler, by S. B. Flagg, G. C. Cook, and F. E. Woodman. 1914. 32 pp., 1 pl., 4 figs.

Technical Paper 50. Metallurgical coke, by A. W. Belden. 1913. 48 pp., 1 pl., 23 figs.

Technical Paper 76. Notes on the sampling and analysis of coal, by A. C. Fieldner. 1914. 59 pp., 6 figs.

Technical Paper 80. Hand-firing soft coal under power-plant boilers, by Henry Kreisinger. 1915. 83 pp., 32 figs.

Technical Paper 97. Saving fuel in heating a house, by L. P. Breckenridge and S. B. Flagg. 1915. 35 pp., 3 figs.

Technical Paper 98. Effect of low-temperature oxidation on the hydrogen in coal and the change of weight of coal in drying, by S. H. Katz and H. C. Porter. 1917. 16 pp., 2 figs.

Technical Paper 123. Notes on the uses of low-grade fuel in Europe, by R. H. Fernald. 1915. 37 pp., 4 pls., 4 figs.

Technical Paper 133. Directions for sampling coal for shipment or delivery, by G. S. Pope. 1917. 15 pp., 1 pl.

Technical Paper 137. Combustion in the fuel bed of hand-fired furnaces, by Henry Kreisinger, F. K. Ovitz, and C. E. Augustine. 1916. 76 pp., 2 pls., 21 figs. 15 cents.

[20]Technical Paper 148. The determination of moisture in coke, by A. C. Fieldner and W. A. Selvig. 1917. 13 pp.

Technical Paper 170. The diffusion of oxygen through stored coal, by S. H. Katz. 1917. 49 pp., 1 pl., 27 figs.

Technical Paper 172. Effects of moisture on the spontaneous heating of stored coal, by S. H. Katz and H. C. Porter. 1917. 25 pp., 1 pl., 8 figs.

PUBLICATIONS THAT MAY BE OBTAINED ONLY THROUGH THE SUPERINTENDENT OF DOCUMENTS.

Bulletin 8. The flow of heat through furnace walls, by W. T. Ray and Henry Kreisinger. 1911. 32 pp., 19 figs. 5 cents.

Bulletin 11. The purchase of coal by the Government under specifications, with analyses of coal delivered for the fiscal year 1908-9, by G. S. Pope. 1910. 80 pp. 10 cents.

Bulletin 13. Résumé of producer-gas investigations, October 1, 1904, to June 30, 1910, by R. H. Fernald and C. D. Smith. 1911. 393 pp., 12 pls., 250 figs. 65 cents.

Bulletin 14. Briquetting tests of lignite at Pittsburgh, Pa., 1908-9, with a chapter on sulphite-pitch binder, by C. L. Wright. 1911. 64 pp., 11 pls., 4 figs. 15 cents.

Bulletin 18. The transmission of heat into steam boilers, by Henry Kreisinger and W. T. Ray. 1912. 180 pp., 78 figs. 20 cents.

Bulletin 21. The significance of drafts in steam-boiler practice, by W. T. Ray and Henry Kreisinger. 64 pp., 26 figs. 10 cents.

Bulletin 22. Analyses of coals in the United States, with descriptions of mine and field samples collected between July 1, 1904, and June 30, 1910, by N. W. Lord, with chapters by J. A. Holmes, F. M. Stanton, A. C. Fieldner, and Samuel Sanford. 1912. Part I, Analyses, pp. 1-321; Part II, Descriptions of samples, pp. 321-1129. 85 cents.

Bulletin 23. Steaming tests of coals and related investigations, September 1, 1904, to December 31, 1908, by L. P. Breckenridge, Henry Kreisinger, and W. T. Ray. 1912. 380 pp., 2 pls., 94 figs. 50 cents.

Bulletin 27. Tests of coal and briquets as fuel for house-heating boilers, by D. T. Randall. 44 pp., 3 pls., 2 figs. 10 cents.

Bulletin 37. Comparative tests of run-of-mine and briquetted coal on locomotives, including torpedo-boat tests, and some foreign specifications for briquetted fuel, by W. F. M. Goss. 1911. 58 pp., 4 pls., 35 figs. 15 cents.

Bulletin 40. The smokeless combustion of coal in boiler furnaces, with a chapter on central heating plants, by D. T. Randall and H. W. Weeks. 1912. 188 pp., 40 figs. 20 cents.

Bulletin 41. Government coal purchases under specifications, with analyses, for the fiscal year 1909-10 by G. S. Pope, with a chapter on the fuel-inspection laboratory of the Bureau of Mines, by J. D. Davis. 1912. 97 pp., 3 pls., 9 figs. 15 cents.

Bulletin 109. Operating details of gas producers, by R. H. Fernald. 1916. 74 pp. 10 cents.

Bulletin 116. Methods of sampling delivered coal, and specifications for the purchase of coal for the Government, by G. S. Pope. 1916. 64 pp., 5 pls., 2 figs. 15 cents.

Technical Paper 20. The slagging type of gas producer, with a brief report of preliminary tests, by C. D. Smith. 1912. 14 pp., 1 pl. 5 cents.

Technical Paper 63. Factors governing the combustion of coal in boiler furnaces; a preliminary report, by J. K. Clement, J. C. W. Frazer, and C. E. Augustine. 1914. 46 pp., 26 figs. 10 cents.

Technical Paper 65. A study of the oxidation of coal, by H. C. Porter. 1914. 30 pp., 12 figs. 5 cents.

Technical Paper 114. Heat transmission through boiler tubes, by Henry Kreisinger and F. K. Ovitz. 1915. 36 pp., 23 figs. 10 cents.

TRANSCRIBER'S NOTES

Added table of contents to HTML version.

Page 5: Added period to the sentence: "If the coal used in the test is to be analyzed, take a sample of from 4 to 6 pounds from each barrow and throw it into a box near the scales.".

Page 11: Changed typo "calcuate" to "calculate."

Page 18: Changed typo "1.1854" to "1.0854", see intersecting columns 184° F and 200 psi.

Page 19: Changed typo "Samuel Sandford" to "Samuel Sanford."

www.ingramcontent.com/pod-product-compliance
Lightning Source LLC
Chambersburg PA
CBHW070745180526
45168CB00004B/1535